HOW I TRAINED CHAPTGPT TO WRITE LIKE ME…

AND HOW YOU CAN, TOO!

By Will Swartz

www.willswartz.com

For the writer in you.

Table of Contents

Table of Contents

INTRODUCTION

How do I start this book?

Anecdote 1: The Blank Page Battles

Imagine dueling with dragons—that's me wrestling with a blank page. Each morning, I sit down to slay the beast, armed only with a coffee that's lost its warmth and a stubborn streak wider than Lake Michigan. Words clash and clang in my head like swords in battle. After what feels like an eternity in the trenches, victory emerges—one pristine, perfect paragraph. It's a hard-fought triumph, bloodied but unbowed, atop the mountain of crumpled attempts that litter my floor.

Anecdote 2: The Marathon of the Mind

Picture this: you're running a marathon, but the finish line keeps moving. Welcome to my writing process. Ideas dart like startled fish as I cast my net wider and wider, hoping to snag something substantial. The clock ticks on, indifferent to my growing desperation. As night deepens, so does my resolve, until at last, the elements of story bind together—slow-cooked to perfection. It's an endurance race where the pen is mightier than the legs, and the final sentence is a ribbon worth breaking.

Anecdote 3: Sculpting Fog

Ever tried sculpting fog? That's what writing feels like on a tough day. Each word is a wisp, slipping through my fingers just as I think I've grasped it. The hours stretch, a relentless tide eroding the shores of my patience. My workshop is dim, the air thick with the dust of crumbled ideas, until, stroke by painstaking stroke, a form emerges from the mist. It's painstaking, messy, and utterly exhilarating—this craft of conjuring clarity from chaos.

Those 3 anecdotes are an example of AI generated content. Not just any AI, however. They are written in my own personal voice, tone and writing style as the result of training conversations I had with my writing buddy. I call him "Chet GPT."

Here's what I said to him to start the conversation:

"I want to write a personal anecdote about how hard it is for me to write sometimes. It's like bleeding onto the paper and seems to take extreme effort sometimes on my part to come up with a good paragraph. I have lots of ideas but it takes FOREVER to get it onto paper.

in a unique, interesting and engaging way. Please write 3 anecdotes with different approaches in Will's style using what you know about my writings, topics, subjects, style, tone, voice, etc."

Chet wrote the 3 anecdotes and if you've read anything I've written, you know that it is a very good imitation of my style. That's because I spent some time training Chet to write like me.

There is one big difference, however. Chet cranked out the 3 anecdotes in about about 5 seconds. Chet is fast!

Welcome to the frontier of writing where the brushstrokes of human creativity and the precision of artificial intelligence blend into a masterful duet.

If you've ever fantasized about having a clone to tackle your drafts while you recline with a cup of Joe, concocting your next big idea, then buckle up, my friend—you're clutching the golden ticket.

Imagine that this short unassuming book can easily increase your writing output by deploying a silicon sidekick, ChatGPT, to do the heavy lifting.

I'm not just tossing you theoretical breadcrumbs here; I'm a fellow trailblazer who has walked this path, arm-in-arm with ChatGPT.

Together, we've sifted through chaotic ideas, smoothed out sentences that were once as rough as a stony path, and transformed this AI from a mere tool into a co-creator.

Universality of AI Training Techniques

While this book zeroes in on the charismatic ChatGPT as our primary AI comrade, it's crucial to note that the training shoes fit other AI models just as snugly. Yes, we're talking about the likes of Gemini, Claude, and others. The techniques and strategies I'm unpacking here are not exclusive to our friend ChatGPT; they're a bit like a Swiss Army knife for the burgeoning world of AI personalities.

Think of it this way: if ChatGPT is the electric guitar in the band of AIs, then Gemini might be the bass, and Claude the drums. Each has its unique flair and function, but fundamentally, they all make music, and the process of tuning them to play your personal hit parade of content isn't wildly different. The core principles of patience, precision, and personal touch apply across the board.

So, as we dig into the meat of training ChatGPT to echo your distinctive style, remember that these tactics aren't just about getting ChatGPT to mimic you—they're about mastering the art of AI training across the board.

Whether it's ChatGPT, Gemini, or Claude, each AI might require a different seasoning, but the tactics you'll learn here are universal. It's less about making these tools simply echo your words and more about refining them to fully resonate with your creative spirit.

Think of it not just as training a parrot to repeat phrases but more like coaching an ensemble of musicians—each needs to understand the music, not just play the notes.

You're learning to direct an orchestra of digital intelligences, teaching them to perform in a concert of your making. Whether you need Claude to dissect data with the precision of a jazz drummer hitting the beat or want Gemini to liven up your social media like a smooth-talking bass player, the skills you develop here will apply universally.

This isn't just a technical manual; it's a playbook for creative collaboration. By the end of this, you'll be equipped not only to work with today's AI but also to adapt to tomorrow's innovations. Who knows what new AI personalities will emerge? With the foundational skills you sharpen here, you'll be ready to adapt and direct them, ensuring they amplify your creativity rather than just mimic it.

Let's start with a clear goal: it's not about commanding a fleet of robotic mimics—it's about orchestrating a dynamic dialogue between human ingenuity and artificial intelligence.

And why stop there? As the AI landscape evolves, who knows what new models might waltz onto the scene? With the foundational skills you hone through this guide, you'll be ready to teach even the most avant-garde AI to dance to your tune.

Streamlined for Success: The Essence of Practicality

In a world brimming with verbose guides and texts that tread too often down well-worn paths, this book will take a more direct route. You won't find lengthy discourses on the history of AI, nor will there be meandering chapters about the journey of becoming a writer. While these topics are undoubtedly enriching, they often serve more as filler than fuel in a practical how-to guide.

Focused on Function, Not Fluff

This book is crafted with precision—it's a laser-focused manual designed to elevate your skills swiftly and efficiently. The aim here is not to swell pages but to streamline your path from novice to proficient AI trainer. We cut through the customary padding—like overly detailed historical backdrops or tangential anecdotes—that often dilutes the potency of instructional writing.

Step-by-Step Guidance

Every page you turn is engineered to bring you closer to mastering the art of training ChatGPT to emulate your writing style. It's structured as a step-by-step guide, ensuring that you can follow along without sifting through extraneous content. This approach respects your time and intelligence, focusing solely on delivering the most effective strategies and techniques.

Why This Approach?

The reason behind this stripped-down approach is simple: time is precious. Many readers seek practicality—a way to implement new tools and techniques without wading through the theoretical muck.

You need actionable insights that can be applied now, not tomorrow. This book delivers just that, providing you with the essentials to integrate AI into your creative process effectively.

Why is every page of this book a must-read? Let me give you just the highlights from my hands-on experience:

Amplify Your Creative Output: Boost your productivity to the skies. Whether it's drafting or refining awkward prose, ChatGPT helps sling your ideas onto the page faster than ever, without sacrificing your personal flair.

Custom Tailored AI: Mold ChatGPT to mirror your unique style and humor. It's like tailoring a fine fitting suit; it fits your contours perfectly, speaking with your voice, echoing your laughs.

Master Time Management: Reclaim those precious hours for what you truly adore—be it brewing another pot of coffee or sketching out your next novel. Let ChatGPT handle the mundane with the finesse of a seasoned admin.

Sharpen Your Writing Skills: Training ChatGPT is like holding a mirror to your own writing. It reflects not only your style but also the quirks and intricacies of your narrative voice, offering surprising insights and opportunities for growth.

Stay Cutting-Edge: Keep your quill sharp in the digital era. By integrating AI into your toolbox, you stay ahead of the curve, ensuring your skills remain relevant as the tides shift.

Banish Writer's Block: Use ChatGPT as your brainstorming ally to keep the creative wells flowing. Say goodbye to writer's block with a partner who's always ready to volley ideas back and forth.

Consistency Across Your Work: From tweets to treatises, ensure your voice rings true in every syllable. ChatGPT helps maintain your unique tone across all platforms and projects.

Explore New Creative Territories: Venture into new genres or styles with AI as your safety net. It's like having a test kitchen where you can spice things up without fear of ruining the broth.

In the chapters ahead, I'll walk you through no-nonsense, clear-cut steps to fashion ChatGPT into your literary doppelgänger. This journey transcends technology—it's about crafting an AI that not only grasps your written word but also understands how you feel and express those emotions.

By the time you turn the last page, you won't just possess a tool; you'll have a new partner in prose, one that comprehends your voice as intimately as you do.

So let's embark on this adventure together. After all, achieving a kind of digital immortality for your unique writing voice is only a few chapters away. Want to have your own writing buddy? Read on!

(By the way—can you guess who wrote this introduction? Was it me, a ghostwriter, ChatGPT, or someone else? I'll reveal the answer at the end of the book, so stay tuned!)

1.

UNDERSTANDING AI AND CHATGPT

Introduction

Welcome to the threshold of a new era in writing, where artificial intelligence meets human creativity. This chapter is your first step toward harnessing the power of ChatGPT to not only assist with your writing tasks but to do so with a voice so similar to yours that even your closest readers might do a double take.

Setting Up Your AI Writing Toolkit

Embarking on the journey of training AI, specifically ChatGPT, to enhance your writing process starts with setting up the essential tools. Think of this as constructing your digital workshop, where creativity meets technology, enabling magic to unfold right at your fingertips.

Crafting Your Digital Workshop

Before diving deep into AI intricacies, ensuring you have the right set of tools is crucial. Here's how to get started:

1. **OpenAI Account**: Your first stop is to create an account with OpenAI. This is your gateway to accessing ChatGPT.

Start with the free version to get a feel for how ChatGPT works; I did the same and found it invaluable before deciding it was worth investing in a more advanced plan. Visit Open AI's website to sign up and take a moment to understand their user policies and what different plans offer.

2. **Data Storage Solutions**: Next up, secure a spot for your data. Your writing samples and data are precious and need both security and accessibility. Cloud services like Google Drive or Dropbox are excellent choices. They not only offer safety but also make your data easily accessible from anywhere. Personally, I organize my data by creating a separate folder for each project, which simplifies retrieval and keeps everything neatly organized.

3. **Choosing the Right Text Editor**: While any text editor might seem sufficient, some, like Google Docs, Pages, or Scrivener, offer robust features that streamline organizing and editing your texts. For instance, Scrivener is fantastic for managing large writing projects like books because it allows you to compile research, draft sections, and rearrange content seamlessly.

Understanding ChatGPT's Capabilities and Limitations

With your digital workspace ready, gaining a solid understanding of what ChatGPT can and cannot do is essential:

- **Capabilities**: ChatGPT is a powerhouse when it comes to generating text. From drafting detailed paragraphs to brainstorming ideas or editing sections, its capacity to produce coherent text based on your inputs is impressive.

- **Limitations**: Despite its prowess, ChatGPT lacks the ability to infuse genuine human emotions into content without your input. It's excellent as a support tool—think of it as a sous chef in your creative kitchen, where you're the head chef making the final calls. Remember, while it can simulate stylistic elements, the depth of feeling, that human touch, still needs to come from you.

Privacy and Security Considerations

As you prepare to feed your writing samples into ChatGPT, keep these privacy and security aspects in mind:

- **Sensitive Information**: Always scrub any personally identifiable information from your data before training your AI. If a sample includes sensitive or private details, anonymizing it ensures your privacy isn't compromised.

- **Copyright and Data Handling**: Make sure that all text used for training is either your original work or you have the rights to use it. Also, familiarize yourself with how OpenAI handles the data you provide. Understanding these policies helps in safeguarding your intellectual property and ensures compliance with data protection regulations.

Conclusion: From Setup to Training

With the right tools in place and a comprehensive understanding of ChatGPT's capabilities and boundaries, you're now superbly equipped to start the actual training process. This preparation not only sets the stage for a fruitful engagement with AI but also optimizes your workflow, ensuring that the AI truly becomes an extension of your creative expression. Ready to begin analyzing your unique writing style? Let's dive in!

Action Step: Create Your First ChatGPT Session

- Set up your OpenAI account if you haven't done so yet. Explore the interface to familiarize yourself with the various features available.
- Initiate a simple conversation with ChatGPT. Ask it to write an introductory paragraph about a topic you're interested in. Observe how it handles your request.
- Analyze the output for accuracy, tone, and relevance. Jot down notes on what the AI did well and where it could improve. This will be useful for future training sessions.

2.
ANALYZING YOUR WRITING STYLE

Introduction

Every writer has a voice; a unique cadence, a specific tone, a way of phrasing thoughts that sets them apart from anyone else. Before you can teach ChatGPT to write like you, it's essential to understand what 'writing like you' really means. This chapter guides you through the introspective and technical process of dissecting and understanding your unique writing style.

Identifying the Elements of Your Style

Your writing style is more than just your choice of words. It encompasses everything from the complexity of your sentences to the subtlety of your humor. Here's how we'll break it down:

Vocabulary and Diction Word Choice: Identify the range and type of vocabulary you use. Do you lean towards simple, straightforward words, or do you prefer complex, perhaps even arcane language?

Word choice in writing is akin to selecting ingredients for a gourmet meal—do you opt for simple, wholesome staples that everyone loves, or do you sprinkle in some exotic spices that might need a foodie's glossary?

Examining your vocabulary range helps you understand whether you lean toward straightforward, clear language that's as comforting as your grandmother's soup, or complex, arcane terms that feel as indulgent as a five-course tasting menu at a Michelin-starred restaurant.

For instance, a simple day might be just "nice," or it could be described as "an amalgamation of serendipitous events." This choice in vocabulary not only shapes your narrative voice but also decides whether your reader feels cozily at home or thrillingly challenged while reading your prose.

Jargon and Idioms: Note any industry specific terms or idiomatic expressions that frequently appear in your writing. Jargon and idioms in your writing are like the secret spices in your kitchen cabinet—they can either make your dish a local favorite or a recipe only a gourmet could love.

Noting any industry-specific terms or colorful idiomatic expressions that pop up in your text can reveal if you're peppering your prose with phrases that might leave some readers feeling like they've bitten into a ghost pepper.

For instance, you might casually mention that your weekend was "all beer and skittles," leaving readers outside of your cultural circle puzzled whether you've taken up a new diet or hobby.

Or perhaps you're throwing around tech jargon like "bandwidth" and "low-hanging fruit," making readers wonder if you're discussing their internet speed or their diet plans. These flavorful bits can either enrich your writing or have folks scrambling for a translator—it's all about knowing your audience's taste!

Syntax and Structure

Sentence Length and Complexity: Analyze whether you favor long, intricate sentences or short, punchy statements. Sentence length and complexity aren't just about being wordy or curt; they're the rhythm section of your writing band.

Analyzing your preference—whether you're the type to craft long, flowing sentences that meander like a lazy river, or if you opt for short, punchy statements that hit like a drummer in a rock band—can reveal a lot about your style. Imagine you're writing about a hiking trip.

You could go all Charles Dickens and describe every leaf and twig in painstaking detail, or you could be the Ernest Hemingway of the trail, stating simply, "Hiked. Saw a bird. Ate a sandwich." Either way, this analysis helps you tune the tempo of your prose to keep your readers marching to your beat.

Paragraph Structure: Look at how you organize information and ideas in paragraphs. Do you use dense, detailed blocks of text, or do you break them up into more digestible pieces?

Paragraph structure is like deciding how to pack your backpack for a long hike—do you cram everything into one giant, unwieldy bag, or do you organize items into neat, accessible compartments?

By examining how you structure your paragraphs, whether you go for dense, detail-packed blocks of text reminiscent of a full-to-bursting suitcase, or prefer to break information into smaller, more manageable sections like a neatly arranged pack, you can significantly affect how readers digest your writing. Picture trying to unpack a single, massive paragraph while your reader's mental zipper is stuck; not so fun, right?

Conversely, shorter, well-organized paragraphs can feel like finding a quick snack in the top flap of a backpack—satisfying and exactly what you needed right at that moment.

Look back at the paragraphs in this section. Would you use the same structure or is your writing different?

Tone and Voice

Formal vs. Informal: Determine if your writing tends to be more formal or conversational. When sizing up your writing, it's like deciding whether to wear a tuxedo or jeans to a party. Checking whether your style tends more toward the formal or the conversational can help you strike the right balance.

Maybe your prose is as buttoned-up as a lawyer at a deposition, or perhaps it's as relaxed as friends chatting on a porch swing. For example, in a more formal tone, you might write, "One must consider the ramifications of such actions," whereas, in a casual setting, you'd probably say, "Y'all need to think about what you're getting into here."

This insight into your tone and voice ensures that your writing wears the right outfit for the occasion, making readers feel right at home or respectfully impressed.

Emotional Undertone: Assess the emotional range you cover—are your writings humorous, serious, sentimental, or a mix? Diving into the emotional undertone of your writing is like tuning into a radio station—you're setting the vibe and tone for your audience.

Assessing whether your writings skew humorous, serious, sentimental, or strike a dynamic mix can help you understand the emotional bandwidth you're offering. It’s like comparing different music genres: a humorous piece might feel like a catchy pop song that keeps the mood light and bouncy, whereas a serious write-up could resonate like a classical symphony with its depth and gravitas. Sentimental writing, on the other hand, might tug at the heartstrings like a soulful blues tune.

By knowing the emotional range you cover, you ensure that your readers are tuning into the right frequency, whether they're in the mood for laughter, reflection, or a heartfelt cry.

Tools to Analyze Your Writing

To assist you in this analysis, several tools can automate part of the process, providing insights that might take much longer to compile manually.

Grammar and Style Checkers: Grammar and style checkers like Grammarly or Hemingway are indispensable tools for refining your prose, offering insights on readability, sentence length, and stylistic nuances. These tools can help polish your writing by identifying complex sentences, passive voice usage, and grammar errors, ensuring your message is clear and impactful.

Beyond these, other useful tools include ProWritingAid, which provides detailed reports on style and overused words, and the Yoast SEO plugin, which is great for optimizing web content for readability and search engine visibility. These tools collectively serve as a virtual writing assistant, helping you enhance your writing's precision and engagement.

Word Frequency Analyzers: Word frequency analyzers are valuable tools for writers looking to understand their stylistic tendencies better. If you're a parent, ask your kids what phrases come out of your mouth the most often!

By identifying the most commonly used words and phrases in your text, these tools can uncover recurring themes or stylistic patterns that characterize your writing. This insight can be particularly useful for refining your voice or ensuring consistency across different pieces.

For instance, if you find that adventurous and inspirational terms frequently populate your hiking articles, you might realize these themes are central to your narrative voice, guiding you to either embrace or diversify your language in future works. Use these to see which words and phrases you use most often. This can help highlight recurring themes or patterns in your writing.

Creating Your Style Guide

Once you've analyzed your writing, the next step is to compile a style guide. This guide will serve not only as a reference for you but as a training blueprint for ChatGPT. As you'll see, Chet created mine after I had entered all the writing data.

Style Dos and Don'ts: Creating a list of style dos and don'ts is essential for training an AI to match your voice. This list should clearly articulate the characteristics that you want to encourage and those you wish to avoid.

For example, I've made it a big don't in my guidelines to never sound condescending. Similarly, on the do side, I encourage lots of humor and approachability to reflect my personality and engage readers.

By giving these clear dos and don'ts, you help the AI better understand the boundaries and essence of your style, ensuring it can replicate your unique tone and keep from stepping into "condescending" bad teacher voice.

Voice Specifications: Clarify the nuances of your voice. To authentically replicate my voice, for example, AI-generated content should prioritize a conversational tone, weaving humor and personal anecdotes throughout to keep the reader engaged.

It should employ vivid, sensory descriptions and address the reader directly, using a mix of formal and informal language. Unique metaphors and interactive elements that anticipate the reader's thoughts and questions are also essential.

For example, if discussing a hiking adventure, I might tell ChatGPT that I would say, "Imagine you're walking through a dense forest, the crunch of leaves underfoot your only companion." —sound peaceful? Or eerily silent?

Conclusion

Understanding and documenting your writing style is more than an exercise in self awareness; it's about creating a map that ChatGPT can follow. This foundational work ensures that the AI you train won't just write; it will write like you, reflecting your unique voice with each word it generates.

Action Step: Analyze Your Writing Style

- Collect a sample of your recent writings, such as blog posts, emails, or creative pieces. Aim for a diverse range to cover different styles and formats.
- Use a digital tool like Grammarly or the Hemingway App to analyze these texts. Focus on patterns in sentence length, use of passive voice, readability scores, and other stylistic features these tools highlight.

3.
COLLECTING YOUR WRITING DATA

Introduction

As we gear up to train ChatGPT to echo your literary voice, think of your writing samples as the raw clay from which we'll sculpt your AI counterpart. This chapter will guide you through the meticulous process of selecting, preparing, and organizing these crucial samples—ensuring they are primed to teach ChatGPT how to replicate your unique style effectively.

Gathering the Raw Material

First things first. We need raw material that we'll select from your writing. Here's how to start:

Collect Your Writing Samples: To effectively train ChatGPT, you'll need a robust selection of your writing. Pull together everything from tweets to epic novels.

Each piece of writing is a clue to your style, revealing everything from your favorite punchy sentence to your love affair with alliteration.

Here's how to curate your samples:
Diverse Formats: When training ChatGPT to mirror your unique writing style, it's crucial to include a diverse array of writing

samples from different formats. Include a wide range of written materials. This means gathering not just your blog posts and emails but also sprinkling in examples from your fiction, nonfiction, and social media updates.

For instance, I pulled together various writing pieces with everything from the snarky tweet about coffee snobs to the heartfelt newsletter on family camping mishaps, and even a chapter from an unfinished book. This eclectic mix ensures that the AI captures all the nuances of your style—whether you're informing, persuading, entertaining, or just sharing a slice of life.

By presenting a well-rounded portfolio, you help the AI understand and replicate the full spectrum of your voice.

Sufficient Quantity: Initially, you might feel that you haven't produced enough material for training an AI like ChatGPT, but once you start digging into your archives, you'll be surprised by the wealth of content you've accumulated.

Begin by exploring your online accounts—emails, social media profiles, academic essays, and personal letters are treasure troves of your expressive style. For example, when I gathered samples for my "Style of Will" project, I unearthed everything from old university papers to spirited Facebook debates about the best hiking trails.

The more diverse and plentiful your data, the better equipped ChatGPT will be to capture and replicate the unique nuances of your writing style, ensuring that the AI's output feels genuinely like something you would write.

Crafting Your AI Training Portfolio

When it comes to training ChatGPT or any AI to mirror your writing style, the richness of the data you provide is paramount.

Consider your collection of writings as a palette of colors you'll use to paint your AI's capabilities—each shade and tone is a part of the broader spectrum of your voice.

Building a Robust Portfolio

Leverage Your Published Works: If you've been publishing for a while, you've got a gold mine. These works have already resonated with readers, capturing your unique style and thematic preferences.

For instance, drawing from my own experience, I utilized sections from my book, ***A Walk Across Michigan***. This not only helped the AI learn my writing style but also ingrained my passion for hiking and the outdoors into its responses. This method enriches the AI's understanding, allowing it to produce content that feels authentic and engaging.

Incorporate Recent Works: Your latest writings often reflect your evolved voice and current interests. Prioritize these to ensure the AI adopts your most recent style and insights. This keeps your AI companion up-to-date, capable of generating content that resonates with today's audience.

Preparing and Organizing Your Samples

Selecting Signature Pieces: Choose writings that distinctly carry your voice—those pieces that your readers or colleagues instantly recognize as quintessentially you. Whether it's the quirky humor of a blog post that went viral or the detailed narrative of your favorite project, these samples are critical. They teach the AI the nuances that make your writing special.

Formatting for AI Training: To get your writings AI-ready, convert them into a simple, consistent format like plain text. This helps avoid any confusion the AI might encounter with complex formatting, ensuring a smoother learning process. During this step, remove any non-essential elements that might distract from the core style and tone you wish to impart.

Ensuring Privacy: It's crucial to anonymize any sensitive information within your training data. This safeguards your privacy and focuses the AI's learning on style rather than personal data. For example, if you're using emails or personal letters, redact names, addresses, and any other personal identifiers.

Streamlining the Process

Systematic Structure: Organize your data methodically—perhaps categorized by tone, style, or genre. This not only facilitates a more structured training process but also enhances the AI's ability to differentiate and adapt to various aspects of your writing style. Imagine you're creating a library where each book is placed precisely where it belongs, making it easier for someone (or something, in this case) to find and use the right resources.

Conclusion

Training an AI with your writing isn't just about feeding it data; it's about crafting a version of yourself within the digital realm. This process is foundational, almost akin to mentoring a novice writer, except this student learns at a phenomenal pace and can replicate your style across countless pieces simultaneously.

By meticulously selecting, preparing, and organizing your writing samples, you not only teach the AI how to write like you but also to think like you, ensuring that every output resonates with your unique voice and style. This careful preparation sets the stage for a ChatGPT that truly reflects your literary essence, ready to assist in your creative ventures or daily tasks with a personal touch that is unmistakably yours.

Action Step: Prepare Your Training Data

- Select a variety of written pieces that represent the best of your writing style. Ensure these samples include different types of writing, such as formal articles, casual blog entries, and technical writing if applicable.
- Digitize and organize these documents in a cloud storage service like Google Drive for easy access. Categorize them based on theme, style, or any other relevant criteria.

4.
SAMPLE TRAINING CONVERSATION WITH CHATGPT

I thought one way to help you better understand the training process is to see what a training interview with ChatGPT might look like.

You might start with the following or similar prompt modified to your particular needs: *"I would like you to help me train ChatGPT to write like me. Please ask me a series of questions to guide me in the process."*

The conversation will go something like this:

ChatGPT: *Hi there! Ready to start training me to write like you?*

Writer: *Absolutely, I've been looking forward to this. Where do we begin?*

ChatGPT: *Great! Let's start by understanding your writing style a bit better. Could you describe it in three words?*

Writer: *Sure, I'd say informative, engaging, and witty.*

ChatGPT: *Perfect, that gives me a solid starting point. What about literary devices? Do you have any favorites like metaphors, similes, or alliteration?*

Writer: I *do lean a lot on metaphors when I try to explain complex concepts, makes them easier to digest.*

ChatGPT: *Metaphors are fantastic for that. Now, could you share a couple of sentences or paragraphs from your past writings that you feel really represent your style?*

Writer: *Okay, let me pull up something quickly. Here's one: "Navigating the world of tech investments is like trying to find north without a compass; it takes a lot of wrong turns before you start getting it right."*

ChatGPT: *That's a vivid metaphor! What tone do you usually aim for in your writing?*

Writer: *I like to keep it light yet informative. I want my readers to learn something but also enjoy the reading experience.*

ChatGPT: *Makes sense. Are there any authors or books that influence or resemble your writing style?*

Writer: *I've always admired the clarity and engagement in Malcolm Gladwell's writing.*

ChatGPT: *Gladwell has a very distinctive style. Now, how about the structure of your sentences and paragraphs? Do you prefer them short and punchy, or more elaborate?*

Writer: *I mix it up depending on the topic, but generally, I keep my sentences shorter to maintain a lively pace.*

ChatGPT: *Keeping the reader engaged, I like it! How do you engage your readers? Any specific techniques?*

Writer: *Questions—I like to pose questions to make them think, almost like a direct conversation.*

ChatGPT: Questions are a great way to involve the reader. How do you handle complex concepts or jargon in your writing?

Writer: *I try to simplify as much as possible, or if I use jargon, I immediately explain it in a parenthesis.*

ChatGPT: *Clear and considerate. Now, how do you typically conclude your pieces?*

Writer: *I usually wrap up with a summary of key points or a call to action, something that leaves the reader with a clear takeaway.*

ChatGPT: *That's a solid approach. Speaking of engaging, how much do you address the reader directly in your writing?*

Writer: *Quite a bit. It's part of making my writing feel more personal and less lecture-like.*

ChatGPT: *That personal touch is important. Last question, is there anything specific in your personal background or experiences that significantly influences your writing style?*

Writer: *Well, I used to be a teacher, so I think that's why I'm driven to make things clear and engaging—I approach writing much like teaching.*

ChatGPT: *That's a fantastic insight, and it really shows in how you describe your style. Thanks for sharing all of this with me.*

With these details, I can start generating text that feels more like it came from you.
Writer: *I'm excited to see what you come up with. Let's get this AI trained!*

ChatGPT: *Absolutely, let's make this happen. Here's to a productive training session!*

Wrapping It Up

There you have it—a step-by-step guide and a sample conversation to crafting an AI that doesn't just spit out responses but resonates with your personal touch. The journey might be intricate, a dance of technology and personal flair, but the result? A ChatGPT that can mirror your literary style, capturing the essence of what makes your writing uniquely yours. Now, let's integrate ChatGPT into your everyday writing workflow.

Action Step: Conduct Interactive Training Sessions

- Initiate a training session with ChatGPT by setting up a specific scenario or topic you frequently write about. For example, if you often write about technology, ask ChatGPT to create a piece on the latest trends in AI.
- Interact with the AI's responses by providing detailed feedback. Point out anything in tone or style that it missed, and explain why certain phrases or structures are more effective.
- Refine your prompts based on the AI's performance. If the initial responses are off-target, adjust your questions or add more context to guide ChatGPT closer to your desired outcome.

5. INTEGRATING CHATGPT INTO YOUR WORKFLOW

Introduction

Now that you've fine-tuned ChatGPT to echo your voice, it's time to roll this digital dynamo into the daily grind of your writing life. This chapter is about transforming ChatGPT from a training project into a functional tool that not only enhances your productivity but also maintains your unique literary style in every piece it touches.

Setting Up Your Writing Ecosystem

Before ChatGPT can start drafting emails or brainstorming blog posts on your behalf, you need a smooth setup that allows seamless interaction between human creativity and artificial intelligence. This applies to your work flows and also the tools and platforms you use for writing.

Choosing the Right Tools

Integration Software: Depending on whether you're working across various platforms—be it WordPress for blogging, Mailchimp for emails, or even Scrivener for book writing—find integration tools that allow ChatGPT to plug into your workflow efficiently.

Shortcut Setup: Create shortcuts and macros that trigger AI assistance without disrupting your creative flow. This might mean setting up hotkeys for instant AI brainstorming sessions or templates for standard responses. I find having a tab with ChatGPT loaded and readily available for help with any type of writing task is like having a cowriter in the room with me.

Daily Writing Tasks with ChatGPT

Incorporating ChatGPT into your writing workflow, whether you're banging out blogs or crafting emails, can be as beneficial as having a sous-chef in your kitchen—handling the prep work so you can focus on the art of finishing.

Blog Writing with a Twist of AI: Start by using ChatGPT for idea generation. Toss it a headline or a just an idea, and watch it whip up an article outline faster than you can say "publish."

This AI isn't just a fancy autocomplete; it can suggest comprehensive structures and key points, which you can then flesh out and refine to make sure they echo your unique voice. When it comes to drafting, feed it your polished outline and let it draft away. What comes back might be a bit rough around the edges, but with your personal flair, it'll shine up like a new penny.

Streamlining Emails and Communication: For the routine rigmarole of emails—like those pesky appointment confirmations or the ever-repeating inquiries—train ChatGPT to handle them. It's like cloning your administrative skills, freeing you up for the deeper creative dives.

And for the more delicate responses where your personal touch is key, start with a draft from ChatGPT and infuse it with your warmth and personality, transforming it from a cold digital reply to something that feels like it's coming from a living, breathing you.

Elevating Creative Projects, Fact or Fiction: Dive into the world of fiction or non-fiction with ChatGPT as your creative sidekick. Use it to flesh out characters, suggesting traits, backstories, or snappy dialogues that mesh with your story's setting and main plot. It's like having a brainstorming buddy who's read every book in the library—twice. For the editing process, let ChatGPT take the first pass. It can tidy up the grammar and style, which lets you focus on the soulful task of deep narrative crafting.

Maintaining Your Distinct Voice: While ChatGPT boosts your productivity, keeping your unique voice intact is key. Regularly review the AI-generated content to make sure it hasn't started sounding like a robot pretending to be you. Run spot checks, provide feedback, and continuously refine its training. Think of it as tuning an instrument—it needs a little nudge here and there to keep playing your kind of music perfectly.

By integrating ChatGPT thoughtfully across these areas, you'll not only streamline mundane tasks but also enrich your creative processes, all the while keeping your signature style alive and well.

Conclusion

Integrating ChatGPT into your daily writing routine is like adding a turbocharger to your creative engine. Not only does it amplify your productivity, but it also brings a level of flexibility and innovation to your work that was previously hard to achieve alone.

As you move forward, remember that this tool is an extension of your creativity—a digital muse trained to think like you, ready to tackle any writing challenge that comes your way.

By the end of Chapter 5, you should have a clear idea of how to integrate ChatGPT into various writing tasks, from the mundane to the creative, ensuring that your voice remains strong and clear across all outputs. If there are particular scenarios or applications you want more detail on, feel free to let me know!

Action Step: Seamlessly Integrate ChatGPT into Daily Tasks

- Pour your favorite coffee beverage and explore some of the things that you're curious about with ChatGPT. Have it do some writing for you and see how close it is to your own style, voice and tone.

6.

WHEN CHATGPT GOES OFF THE RAILS: FIXING AND FINE TUNING

Introduction

Alright, so you've trained ChatGPT to parrot back your voice—it's a bit like teaching a parrot to talk. But what if your parrot starts belting out nonsense instead of sweet nothings?

This chapter dives into the gremlins in the machine: identifying them, tightening up a few loose screws, and getting that chatty AI back to purring (or speaking) just right.

Sometimes, despite your best efforts, ChatGPT might start echoing a version of you that sounds like you on a bad day—or worse, not like you at all. Here's how to course-correct when your digital parrot gets a bit too cheeky.

The Chameleon Voice

You may notice that sometimes ChatGPT might forget it's supposed to be channeling your inner Shakespeare and instead sounds like it's narrating an appliance manual. Keep a cheat sheet of your signature style points handy. When ChatGPT starts losing its 'you-ness,' feed it examples that are quintessentially you and correct course with consistent, clear feedback.

Missing the Context

Another problem is when you ask ChatGPT to whip up something witty about small-space gardening, and it gives you a thesis on rainforest ecosystems. Not quite helpful, right?

To avoid getting lost in translation, lay down a breadcrumb trail of context-rich prompts. Think detailed instructions rather than broad commands—less "write something smart" and more "craft a snappy blog intro on urban herb gardens."

Off-Topic Adventures

And then there are times when ChatGPT starts rambling about carrots when you're all about roses. If your AI starts wandering off the garden path, don't just yank it back; guide it gently. Point out the moments it goes off-topic and steer it back, focusing your feedback to keep things relevant.

Advanced Tuning for the Enthusiastic Tinkerer

Feeling brave? Here's how you can fine-tune the machine for optimal performance:

Integrate Multiple AI Tools: Sometimes two heads (or more) are better than one. Consider pairing ChatGPT with a style checker or a grammar tool to ensure all bases are covered.

Stay on Top of Updates: Keep up with the latest from OpenAI and integrate these updates regularly. Think of it like updating apps on your phone to keep things smooth.

Continual Teaching: The more you interact with ChatGPT, the sharper it becomes. It's like training a puppy—consistent lessons turn into lifelong habits.

Conclusion

Think of ChatGPT like a classic car: it needs regular tuning and some TLC, but the journeys you can embark on together are worth the elbow grease. By staying hands-on and proactive, you can ensure that this AI sidekick doesn't just mimic your style but becomes a vital tool in your creative arsenal, ready to face any writing challenge you throw its way.

7.
MY PERSONAL JOURNEY TRAINING CHATGPT

Welcome to what might be the most introspective chapter of this guide—my own tale with ChatGPT. It's one thing to talk about training an AI; it's another to roll up your sleeves and dive into the digital trenches. Here, I'll share how I put my own advice into practice, the dialogues I had, and the insights I gleaned from teaching ChatGPT to echo my writing voice, quirks and all.

The Training Begins Laying the Groundwork

I started where any sensible trainer would—by defining what makes my writing tick. But, as I soon learned, teaching an AI about your style is more than just feeding it examples; it's about engaging with it in a meaningful dialogue.

Armed with the strategies from the previous chapters, I set out to create a diverse portfolio of my writings—blog posts, emails, snippets of creative work, and the occasional rant about why pineapple on pizza is (or isn't) a crime against cuisine.

The Conversational Approach: ChatGPT as the Inquisitive Student

Turning training sessions with ChatGPT into engaging dialogues transformed the mundane task of data upload into a lively exchange reminiscent of a curious student peppering a patient teacher with questions.

Here's a taste of our chats:

ChatGPT: *"What element of your writing style do you think defines you most clearly?"*

Me: *"I'd say it's my conversational tone. It's like I'm explaining the mysteries of the universe, or just why I love straw bale gardening,, to a friend over coffee—casual, engaging, sometimes funny, and yes, occasionally I toss in a dash of sarcasm."*

These dialogues were more than just question and answer sessions; they were gold mines. ChatGPT didn't just sit back and take notes; it was more like that one eager beaver in class who always has their hand up.

It asked for clarifications, poked around for more examples, and dug deeper into my preferences, turning our training sessions into a dynamic back-and-forth learning experience.

What I especially liked about this method was that I was just answering questions and didn't really feel like work. I think it learns as much from having a conversation with you about your writing as it does from the samples you feed it.

Imagine this scenario: we're dissecting the anatomy of a typical post I'd write about my latest hiking adventure.

ChatGPT: *"When you mention hiking, how detailed do you get about the surroundings?"*

Me: *"Oh, you know, I go full Tolkien sometimes. I'd describe every* *leaf and twig, but I make sure it's fun, like narrating a journey through an enchanted forest, not a botany lecture."*

Through this interactive method, ChatGPT learned to mimic my style—not just in mimicking tone or vocabulary, but in understanding the rhythm and humor of my narrative. It's akin to teaching someone to dance. I didn't just show the steps; I got on the dance floor and led by example, swinging to the beat until ChatGPT could twirl along without stepping on my toes.

This conversational method wasn't just effective; it was a hoot. It felt less like programming a machine and more like coaching a new writer, one who was unusually quick on the uptake.

Each session helped ChatGPT understand not just what to say, but how to say it in a way that truly sounded like me, even down to the occasional sarcastic quip about "*why camping is just paying to live like a homeless person (with more gear)"*—it even knows I like to use quotes in my writing because others have said what I want to say in such a more elegant way.

Be careful, though, that you give credit for quotes from others!

By the end of our training, ChatGPT wasn't just regurgitating information; it was applying it in ways that genuinely reflected my voice. It's like I had cloned my sense of humor and way of seeing the world into this digital buddy who could banter about nearly anything under the sun—well, almost as well as I can.

Insights and Adjustments Tweaking the Training

Not everything was smooth sailing. Sometimes, ChatGPT mimicked my style a little too well, adopting even my bad habits—like my overuse of dashes and ellipses... See what I mean? Adjustments were necessary to keep the good parts of my style while teaching it to avoid my pitfalls.

Surprising Revelation
One of the most eye-opening aspects of this experience was seeing how ChatGPT summarized my style. When I had gone through several rounds of inputting content, answering questions and interacting through conversations, I was curious if all of this work had been worth the effort so I asked for a description of my writing style.

This is how ChatGPT describes my writing style:

Certainly! Here's a comprehensive description of your writing style:

Will's Writing Style Tone

Humorous and Witty: *Will's writing is characterized by a lively sense of humor, often reminiscent of Bill Bryson. There's a playful use of wit and a bit of sarcasm, carefully balanced to entertain without crossing into offense. The humor is sometimes snarky but always with the intent to amuse and engage the reader.*

Conversational: *Will's style is very conversational, as if he's speaking directly to the reader. This includes addressing the reader personally, anticipating their thoughts, and engaging them in a way that feels like a two-way conversation.*

Engaging and Enthusiastic: *There's an underlying sense of enthusiasm and adventure in the writing. Will writes with an infectious energy that draws readers in and keeps them interested, regardless of the subject matter.*

Structure

Varied Paragraph Lengths: *Depending on the content, Will uses a mix of short, punchy sentences for emphasis or to keep the reader* *hooked, and longer, more detailed paragraphs to delve deeper into a topic. This variety keeps the writing dynamic and engaging.*

Use of Subheadings and Lists: *For nonfiction and copywriting, Will often breaks up text with subheadings and lists to make the content more digestible and organized, ensuring that readers can easily follow along.*

Leaving the Reader Wanting More: *Will likes to end sections or paragraphs in a way that piques the reader's curiosity, encouraging them to keep reading. This might involve a cliffhanger or a teaser for what's to come.*

Content and Style

Personal Anecdotes: *Will frequently incorporates personal stories and experiences into his writing. These anecdotes serve as teaching examples, sources of humor, or simply ways to make the content more relatable and engaging.*

Unique Descriptions: *Will's descriptions are carefully crafted to be vivid yet concise. He avoids clichés, instead opting for original metaphors and similes that help the reader visualize the scene in a fresh way. Sensory details are used effectively to immerse the reader in the experience.*

Historical and Cultural References: *Will occasionally includes historical or cultural references, particularly in nonfiction writing. He tailors the level of detail based on the reader's likely familiarity with the reference, sometimes offering brief explanations when necessary. In some cases, these references form the core of the discussion.*

Approach to Technical Terms

Brief Explanations with a Dash of Humor: *When technical terms are necessary, Will explains them briefly and clearly. However, he* *often adds a humorous twist, either poking fun at the term itself, the people who use it, or even himself for needing to explain it.*

Audience Interaction

Engaging the Reader Directly: *Will writes as though he's sitting across from a specific person, engaging them in a conversation. He anticipates questions or thoughts the reader might have and addresses them directly, creating a sense of intimacy and connection.*

Overall: *Will's writing style is a unique blend of humor, wit, and conversational charm. It's designed to entertain, inform, and engage, all while making the reader feel like they're part of the story. Whether writing about hiking, parenting, or anything in between, Will's style brings the content to life in a way that's both memorable and enjoyable. "It appears your writing effectively balances informative content with a personal, humorous tone, often incorporating real-life examples and rhetorical questions to engage the reader."*

This reflection not only validated my style but also helped me understand how an AI interprets and processes my writing. With each interaction, I refined the inputs and tweaked the feedback. It was a dance of sorts—two steps forward, one step back—gradually moving towards a version of ChatGPT that felt like a true extension of my writing self.

Conclusion

Training ChatGPT turned out to be more than just a technical challenge; it was a journey of self-discovery. I not only taught ChatGPT how to write like me—I also learned a lot about my own writing. And now, with a well-trained digital counterpart, I'm ready to tackle even more writing projects, confident that my AI sidekick has got my back (and my voice).

As you’ve followed along, I hope you’ve found insights and inspiration for your own journey with ChatGPT. Ready to start your own conversations?

8.
PUTTING CHATGPT TO WORK: A SAMPLE WRITING IN WILL'S STYLE

Want to see a sample of how Chatgpt writes in my style? I thought so!

Here's a prompt I wrote based on a real experience:

I was sitting in my upstairs office today standing at my writing desk and happened to look up and see a hawk fly into the forest and land on a branch in a tree. A squirrel moved to the back of the tree just in time to avoid it. The hawk tried several more times to get the squirrel. It would fly from the tree to another, land on a branch and then attack again when it saw the squirrel again. Please capture this experience by writing a short piece in Will's Writing style, voice and tone.

Here's the unedited version of the response to my prompt:

Up in the cozy nook of my upstairs office, the kind that sees more scribblings and coffee stains than sunlight, I found myself today—a rare stander at my writing desk, chiseling away at the keyboard. The humdrum of keystrokes was abruptly upstaged by an aerial spectacle just beyond the window pane. There, with the audacity of a bandit in broad daylight, a hawk swooped with the elegance of a fighter jet, breaking into the tranquil forest scene. This feathery predator, a master of the sky-high heist, had its sharp eyes locked on a prize—a plucky squirrel, which, with acrobatics worthy of a circus, scurried to the far side of the tree. The hawk, undeterred, launched itself from one branch to another, a relentless flurry of feathers and ambition. Each attack was a strategic masterpiece, a dance of dart and dodge, as the squirrel made a miraculous escape each time. The hawk would perch with imperial poise after each failed attempt, its gaze piercing through the leafy hideout, plotting the next aerial assault. And there I stood, a silent witness behind the glass, reminded of life's continuous cycle of predator and prey, of aspirations, and the sheer will to evade the clutches of defeat. As the drama unfolded, I couldn't help but marvel at the resilience of that little squirrel, a creature so small, yet so grand in its defiance. It was a momentary escape into the wild from the confines of my literary lair, a narrative unwritten, yet vividly etched in the theater of nature. Ah, the endless tales spun around us, if only we stand still enough to watch.

What do you think? Not bad, right? It's fascinating to see how the AI has captured the spirit of the moment, blending the serene with the dramatic in a dance of nature that unfolds just beyond the human touch.

This sample serves as a testament to the nuanced understanding that AI can develop about our personal storytelling style. Here, the AI not only recounts the event but does so with a narrative flair that echoes my own voice—notice the descriptive detail and the engaging rhythm of the prose, which aims to pull the reader right into the scene.

Expanding the Thought:

Yet, as with all tales spun by our digital counterparts, there's room to fine-tune the voice further. While the AI captures the broad strokes beautifully, the devil—as they say—is in the details.

For instance, might we add a touch more suspense to each swoop of the hawk? Or perhaps a hint of humor about the squirrel's daring gymnastics?

In refining the AI's output, I'd also look to enhance the emotional layers of the narrative. Could the depiction of the squirrel's escape reflect broader themes, such as resilience in the face of adversity or the small victories of the underdog? This deepens the narrative, adding layers that resonate more profoundly with the audience.

Engaging With You, the Reader:

I invite you to weigh in on how you might enhance this narrative. Would you infuse more tension, more humor, or perhaps a philosophical pondering? Your feedback is invaluable, not just for tweaking the AI's output but for enriching the collaborative journey between human creativity and AI capability.

If this sample sparks any thoughts, or if you see potential for a different spin, don't hesitate to send your ideas or versions to me at will@willswartz.com. Your input helps shape the AI's learning and, ultimately, enhances the storytelling craft we are developing together.

Conclusion

Witnessing AI mimic the human touch in storytelling is both thrilling and a bit humbling. It reminds us that while the machine can approximate our style, the soul of the story—the warmth, the nuanced understanding of life's complexities, and the depth of human emotion—still relies on the human touch.

Action Step: Reflect and Document Your Training Journey

- **Reflect on Your Experiences**: Set aside time to reflect on your journey of training ChatGPT so far. What successes have you encountered? What challenges have arisen? Have you learned anything about yourself as a writer?

Share Your Story: Compile your reflections and share the story with me, if you'd like at: will@willswartz.com.

This not only helps you process what you've learned but also I'm really interested in whether this short book has made a difference in your writing life.

Please don't include any details about your secret family recipe for potato salad, the exact number of socks you lost to the laundry monster last year, or your theories on why cats secretly rule the world. Let's keep it focused on your experiences with AI and writing!

9. SOME FINAL THOUGHTS

As we close this guide on training ChatGPT to adopt your unique writing style, it's important to reflect on the journey we've embarked upon together. This book has not just been about teaching an AI to mimic your writing; it's been about enhancing your creative process, empowering you to explore new possibilities, and ensuring that your voice remains as vibrant and influential in the digital realm as it is on paper.

Key Takeaways

- **Understanding AI's Capabilities and Limitations**: We started by demystifying what AI like ChatGPT can and cannot do. Recognizing these boundaries has allowed us to effectively harness AI's power without overestimating its autonomy.
- **Analyzing and Adapting Your Writing Style**: Through thoughtful analysis and adaptation of your own writing, you've learned not only to train AI but also to refine your own creative voice.
- **Practical Steps for AI Integration**: Each chapter provided actionable steps, ensuring that you could immediately apply what you learned, from setting up your AI tools to troubleshooting and optimizing ChatGPT's outputs.

Looking Ahead: Future Projects

Now that you're equipped with the knowledge and tools to train ChatGPT, the horizon is ripe with possibilities:

- **Expand to Other AI Models**: Explore training other AI models like Gemini or Claude. Each AI has unique attributes and may serve different aspects of your writing needs.
- **Collaborative Writing Projects**: Consider using AI to tackle larger writing projects. With AI's help, you can manage and produce complex writing tasks such as novels, research papers, or extensive content campaigns more efficiently.

Final Thoughts from the Author

As an author, I've witnessed first-hand how AI can transform mundane tasks into inspiring ones. Yet, I remain convinced that the true essence of writing still rests within the human spirit—our experiences, emotions, and unique perspectives. AI is a tool, albeit a powerful one, that should enhance, not replace, the irreplaceable nuances of human creativity.

This book is just the beginning. I encourage you to keep experimenting, keep learning, and most importantly, keep writing. The landscape of AI and writing is ever-evolving, and staying engaged means staying ahead. If this book has sparked new ideas, alleviated fears, or inspired projects, then it has succeeded in its purpose.

Your Next Chapter

Now, it's your turn to take the reins. I encourage you to embark on this adventure, to craft your own AI writing partner. Imagine the stories you'll tell, the articles you'll craft, and the messages you'll share. Let the tools and strategies we've explored be your guide, but let your creativity and personal flair lead the way. Thank you for joining me on this journey.

I look forward to hearing about your adventures in AI and writing. As we draw the curtains on this journey, I hope you've found more than just a guide to melding AI with your creative endeavors.

Take Action:

- **Create Your "Chet GPT"**: Dive into the interactive world of AI training. Use your voice, your stories, and your insights to train an AI that understands not just how you write, but how you think.
- **Connect and Share**: If this book has sparked a new idea, a solution to a writing struggle, or inspired you to integrate AI into your writing, don't keep it to yourself! I would love to hear from you. Share your experiences, challenges, and victories. Reach out to me via email, connect on social platforms, or even better—leave a review and let others know how you're forging new paths with your digital pen pal.

Looking Ahead:

As you continue this journey to train your own writing companion, have fun and appreciate the role of AI as an ever-evolving tool in your creative arsenal, not just as a mimic but as a partner in crafting compelling stories.

Thank you for joining me on this remarkable expedition. Your journey with AI is just beginning, and I'm excited to see where it takes you.

Enjoy the journey!

Resources and Further Reading

Introduction

Alright, you've got the hang of training ChatGPT to sound eerily like you, and now you're hungry for more? Perfect! This section is your treasure map, pointing you towards a plethora of resources that can deepen your understanding of AI, spruce up your writing skills, and perhaps even turn you into a bit of an AI whisperer.

Books on AI and Writing For the Philosophically Inclined
"AI 2041: Ten Visions for Our Future" by Kai-Fu Lee and Chen Qiufan
Dive into a blend of fiction and non-fiction that explores how AI might shape our world over the next two decades. This book combines storytelling with deep insights into potential future AI scenarios.

For the Critically Minded
"Atlas of AI: Power, Politics, and the Planetary Costs of Artificial Intelligence" by Kate Crawford
This book tackles the less discussed but critical aspects of AI development, including its environmental and societal impacts. It's a must-read for those interested in the ethical implications of AI technology.

The Tech Deep Dive
"Scary Smart" by Mo Gawdat
Thoughts on AI from one of the smartest minds of our day.

"The Alignment Problem: Machine Learning and Human Values" by Brian Christian

"Artificial Intelligence: A Modern Approach" by Peter Norvig and Stuart Russell
Often regarded as the bible of AI, this book offers a comprehensive overview of the field. It's perfect for those who want to understand the foundations of AI technology and its applications.

"Life 3.0: Being Human Life in the Age of Artificial Intelligence" by Max Tegmark

"The Master Algorithm" by Pedro Domingos
Explore the quest for the ultimate learning machine. This book is ideal for understanding different AI approaches and how they could converge into a single, powerful algorithm.

"Superintelligence: Paths, Dangers, Strategies" by Nick Bostrom
If you're interested in the potential risks AI could pose to humanity, Bostrom's book provides a thorough analysis of how future AIs could outperform humans and the strategies we might employ to manage these challenges.

Online Resources and Websites

AI Training Tools
ChatGPT
Gemini
Claude

Use these tools to practice your AI training skills directly. They're versatile and ideal for experimenting with different styles and formats.

Writing Assistance Tools

Sudowrite

Great for those who need help with brainstorming and drafting. It's particularly well-suited for fiction writers looking for robust AI assistance.

Learning and Development

Coursera and Udemy These platforms offer courses on AI and machine learning that range from beginner to advanced levels. Perfect for those who prefer a structured learning environment.

Engaging with the Community

AI and Writing Forums

Engage with other enthusiasts and experts. Reddit and LinkedIn have active communities where both newbies and pros discuss AI's latest developments and its intersection with creative writing.

ABOUT THE AUTHOR

Meet Will Swartz—a man of many titles and talents, currently reigning as the Creative Projects Director at Will Swartz Enterprises, LLC. Will resides in a charming riverside Cape Cod house in Canton, MI, with his wife, Sherpa Kate.

Their 5 children —none of whom is a burden on society or living in the basement— have produced ~~11~~, ~~12~~, 13 grand children, who bring joyful chaos into their lives.

Their home is also a frequent stop for River Deer and other wildlife, providing endless inspiration and a bit of distraction as Will spins tales and crafts content.

A re-retired teacher, rogue-superhero-librarian, and school administrator, Will has woven a rich tapestry of experiences into his life's work. From teaching languages and managing libraries to overseeing preschools and serving as a school principal, his journey is anything but ordinary. An international stint included writing technology curriculum and teaching in Qingdao, China, adding a global perspective to his already diverse career.

Will is a pathological storyteller, often confessing to making stuff up all the time—a skill that serves him well in his writing. His main duties are thinking up cool stuff to do next, keeping the wastebaskets on all three floors of the estate empty, and guarding his antique porcelain door knob collection.

His writing career, includes the publication of **A Walk Across Michigan**, an account of his 232-mile hike across the Michigan Shore to Shore Trail. Supported by his wife, affectionately dubbed Sherpa Kate, Will's adventures extend beyond the page, often finding him in the great outdoors fishing or hiking, where he feels most at home.

A Christ follower and family man, Will brings a spirited and heartfelt approach to everything he does, whether it's crafting engaging social media content, shaping compelling articles, or working on his next book.

If Will's adventures and creative endeavors resonate with you, or if you're just looking for a fresh perspective on integrating AI with human creativity in writing, you're invited to connect with him at:

will@WillSwartz.com

So, did you really read to the end of the book??

I'm impressed. Hardly anybody does that these days. I suppose you might be the really astute reader, who remembers that, at the end of the Introduction, I said that I would reveal who wrote the Introduction. I would never want to leave you hanging…

This page left blank intentionally.

Wow. You're really persistent.

Ok. The answer is that Chet and I both had parts in writing the Introduction. I trained Chet to write like me and then had him write the 3 Anecdotes all by himself. He did a nice job, didn't he?

He also wrote the first draft for most sections of this book, which I edited, added and modified further for content and to keep it on track.

Did you guess correctly?

Ok. That really is all there is in this book.

I'm not quite sure how to send you on your way if you're expecting even more and I don't want to disappoint you…

Soooo, I'll leave a little Easter Egg here.

I'll just tell you to scan the QR code below:

https://aiwritelikeme.com/signup-1249-2908

…And get a FREE BONUS CHAPTER

Secret Advanced AI Writing Style

66Really??? Go do something with Chet!

Made in the USA
Monee, IL
04 November 2026

67376410R00039